99-CTS-2ET

ULTRASONIC DESTRUCTION OF SURFACTANTS: APPLICATION TO INDUSTRIAL WASTEWATERS

by:

Linda K. Weavers

J. Aaron Frim

Department of Civil and Environmental Engineering and Geodetic Science, The Ohio State University

Gim Yang Pee

James F. Rathman

Department of Chemical Engineering, The Ohio State University

2003

Water Environment Research Foundation

The Water Environment Research Foundation, a not-for-profit organization, funds and manages water quality research for its subscribers through a diverse public-private partnership between municipal utilities, corporations, academia, industry, and the federal government. WERF subscribers include municipal and regional water and wastewater utilities, industrial corporations, environmental engineering firms, and others that share a commitment to cost-effective water quality solutions. WERF is dedicated to advancing science and technology addressing water quality issues as they impact water resources, the atmosphere, the lands, and quality of life.

For more information, contact:
Water Environment Research Foundation
635 Slaters Lane, Suite 300
Alexandria, VA 22314-1177
Tel: (703) 684-2470
Fax: (703) 299-0742
www.werf.org
werf@werf.org

This report was co-published by the following organization. For non-subscriber sales information, contact:

IWA Publishing
Alliance House, 12 Caxton Street
London SW1H 0QS, United Kingdom
Tel: +44 (0) 20 7654 5500
Fax: +44 (0) 20 7654 5555
www.iwapublishing.com
publications@iwap.co.uk

Library of Congress Catalog Card Number: 2003106625

WERF ISBN: 978-1-893664-81-4   IWAP ISBN: 978-1-84339-636-9

The Ohio State University

ACKNOWLEDGMENTS

The Naval Facilities Engineering Service Center (NFESC) and the assistance of Katherine Ford at NFESC are gratefully acknowledged for funding and stimulating discussions. In addition, the authors would like to thank Limei Yang for experimental assistance.

Report Preparation

Principal Investigator:
Linda K. Weavers, Ph.D., P.E.
The Ohio State University

Project Team:
J. Aaron Frim, M.S.
Gim Yang Pee, M.S.
James F. Rathman, Ph.D.
The Ohio State University

Project Subcommittee

Sandra Kaiser, Ph.D.
Procter and Gamble

Dana Schnobrich, M.S.
3M Company

Water Environment Research Foundation Staff

Director of Research: Daniel M. Woltering, Ph.D.
Project Manager: Roy Ramani, Ph.D.

ABSTRACT & BENEFITS

Abstract:

This research focused on the use of sonication to destroy surfactants and surface tension properties in industrial wastewaters that affect traditional waste treatment processes. We have investigated the sonochemical destruction of surfactants and a chelating agent to understand the release of metals from surfactants during sonication. In addition, the effects of physical properties of surfactants and the effect of ultrasonic frequency were investigated to gain an understanding of the factors affecting degradation. An anionic surfactant, octylbenzene sulfonic acid (OBS), and a nonionic surfactant, t-octylphenoxy polyethoxyethanol (OPE), were chosen as model surfactants. 4-nitrophenol (PNP) and ethylbenzene sulfonate (EBS) were chosen as model non-surfactant compounds. Ethylenediaminetetraacetic acid (EDTA) was used as a model chelating compound.

Sonochemical degradation of surfactants was observed to be more effective than non-surfactant compounds. In addition, as the concentration is increased, the degradation rate constant does not decrease as significantly as with non-surfactant compounds in the near-field acoustical processor (NAP) reactor. In fact, the total number of molecules degraded increases with concentration. The degradation of metal complexes is not as effective as in the absence of the metal. However, this is likely an artifact of the model complexing agent used. Surfactant metal complexes are expected to be faster as they will accumulate at the hot bubble interface, significantly increasing ligand exchange kinetics and thus degradation of the complex.

Benefits:

It has been known that a surfactant may act to chelate metals in solution, possibly causing difficulty with the treatment of metals in wastewater processes. The current research involves the use of sonication for partial or full degradation of surfactants resulting in the release of metals bound to surfactants that in turn may result in a significant cost savings to treatment plants. Even though biodegradable surfactants can be treated by biological treatment processes, this treatment often is land-intensive, expensive, and requires careful maintenance.

- This research on sonication highlights methodologies to understand the effects of physical properties of surfactants and the effect of ultrasonic frequency on the factors affecting destruction of surfactants.
- These methodologies provide a good application potential in the understanding of sonochemical properties and treatment of "improved" detergents which may not be biodegradable.

Keywords: Alkylphenol ethoxylates, sonication, degradation, chromium, ethylenediaminetetraacetic acid.

TABLE OF CONTENTS

LIST OF TABLES

LIST OF FIGURES

LIST OF ACRONYMS

APE	alkylphenol ethoxylates
DTAC	dodeycltrimethlyammonium chloride
EBS	ethylbenzene sulfonate
EDTA	ethylenediaminetetraacetic acid
HTA	hydraxyterephthalic acid
IC	ion chromatography
IPC-AES	inductively coupled plasma atomic emission spectroscopy
NAP	near field acoustical processor
NUWC	Naval Underwater Warfare Center
OBS	octylbenzene sulfonic acid
OPE	t-octylphenoxy polyethoxyethanol
PNP	4-nitrophenol
SDS	sodium dodecylsulfate
TA	terephthalate
TOC	total organic carbon

Executive Summary

Surfactants are major anthropogenic components in wastewaters in the United States. A surfactant may act to chelate metals in solution, possibly causing difficulty with the treatment of metals and polymer flocculation processes in wastewater treatment. Thus, concern over the environmental impact of surfactants is growing and methods to treat surfactants need to be developed. Destruction of biodegradable surfactants typically occurs by biological treatment processes in wastewater treatment plants. For some applications it is not feasible to select a surfactant that is biodegradable, and it is possible that its toxicity properties are not well characterized or understood.

Sonolysis is a result of the formation of cavitation bubbles when sound waves in the range of 20 kHz to several MHz are applied to a liquid medium converting sound energy into heat for chemical reaction. At a critical radius, the cavitation bubble implodes, producing high temperatures and pressures resulting in the formation of H atom and OH radicals. Chemical reactions can either occur through thermolysis within the bubble cavity and at the interface or through OH radical attack at the interface and in the bulk solution.

This research focused on the use of sonication to destroy surfactants and surface tension properties in industrial wastewaters that affect traditional waste treatment processes. We investigated the sonochemical destruction of surfactants and a chelating agent to understand the release of metals from surfactants during sonication. In addition, the effects of physical properties of surfactants and the effect of ultrasonic frequency were investigated to gain an understanding of the factors affecting degradation. Successful partial or total destruction of surfactants resulting in the release of metals bound to surfactants may result in a significant cost savings to treatment plants.

An anionic surfactant, octylbenzene sulfonic acid (OBS) and a nonionic surfactant t-octylphenoxy polyethoxyethanol (OPE) were chosen as model surfactants. 4-nitrophenol (PNP) and ethylbenzene sulfonate (EBS) were chosen as model non-surfactant compounds. Ethylenediaminetetraacetic acid (EDTA) was used as a model chelating compound Three ultrasonic reactors were used for sonochemical experiments. The first was a near-field acoustical processor (NAP) with two stainless steel parallel plates operating at 16 kHz on one plate and 20 kHz on the other. The NAP was run in a continuous circulation mode. The second reactor was a 20 kHz ultrasonic probe operating under batch conditions. The other reactor was a high frequency laboratory reactor operating under batch conditions at 354 kHz.

First, we investigated the effect of the surface-active properties of 4-octylbenzene sulfonic acid (OBS) and t-octylphenoxy polyethoxyethanol (OPE) on their degradation rate constants compared to non-surfactant compounds (PNP and EBS). The degradation of OPE and OBS were enhanced by a factor of 2 and 3 compared to PNP and EBS in a 20 kHz low power intensity reactor (NAP). Enhanced degradation of surfactants was not observed in a high power intensity reactor at 20 kHz (probe). Results demonstrate that degradation rate constants are dependent on the flux of contaminants to the cavitation bubble surface, which is a function of diffusivity and surface excess. Enhanced degradation of surfactants in the NAP was attributed to the accumulation of surfactants on cavitation bubble surfaces resulting in localization of contaminants at the site of highest [•OH]. The lack of enhanced degradation of surfactants in the probe was attributed to the higher power intensity resulting in rapid bubble growth cycles with insufficient time for surfactants to diffuse to and accumulate on the cavitation bubble surface.

It has been observed in sonochemical systems that for increasing concentration of a contaminant, the pseudo first-order rate constant decreases. Thus, the effect of concentration was investigated with surfactants to determine if ultrasound would be effective under a wide range of conditions possibly seen in an industrial wastewater treatment plant. The probe shows an exponential decrease in rate constant with increasing concentration. However, in the NAP, at concentrations up to 0.1 mM, rate constants for OPE do not have the same trend. Above this concentration, the decrease in rate constant is significantly less than in the probe. We believe this phenomenon is unique to surfactants; the longer lifetime of the cavitation bubbles in the NAP allows the surface tension reduction phenomena associated with surfactants to enhance the rate of reaction. Hence, there is a balancing effect between the concentration of the surfactant and the surface tension reduction observed for that concentration.

Finally, the degradation of EDTA, a non-surface active chelating substance, was achieved at two different frequencies, 20 kHz and 354 kHz. It was demonstrated in this study that the system operating at 354 kHz was more effective in degrading EDTA and metal-EDTA complexes than the system at 20 kHz. The increased production of hydroxyl radical and the increased degradation rate of the target organic species at 354 kHz were achieved using a lower power input than that for the 20-kHz system. The presence of a chelated-metal decreased the degradation rate constant of EDTA at both frequencies. Two metals were studied with very different ligand exchange kinetics, a property that appears to control the sonochemical and OH radical degradation rate constant of the complexes. The Cr(III)-EDTA complex, representing a thermodynamically and kinetically more stable complex, degraded at a dramatically lower level than both the Cu(II)-EDTA complex and free EDTA.

CHAPTER 1.0

Motivation and Introduction

1.1 Motivation

Approximately 500,000 tons of alkylphenol ethoxylates (APE) are used annually worldwide (Renner, 1997), making APEs the most widely used surfactants. In 1984, it was discovered that APE breakdown products are more toxic to aquatic organisms than APE itself; hence, APE was banned or restricted in many European countries. Recent evidence also suggests that APE breakdown products are weakly estrogenic. Surfactants are major anthropogenic components in wastewaters in the United States. In addition, several surfactants have been measured in receiving waters of wastewater (Renner, 1997). Thus, concern over the environmental impact of surfactants is growing.

In addition, new industrial-grade surfactants and detergents are created continuously to replace existing detergents and solvents in industrial cleaning processes. These "improved" detergents often are more difficult to treat and result in fouling of industrial wastewater treatment processes. For example, at the Naval Underwater Warfare Center in Keyport, Washington, the use of a variety of non-ionic and anionic surfactants results in failure of the chromium reduction and polymer flocculation processes. The EPA-suggested solution of segregating waste streams before treatment does not work in this case because the chromium and surfactants are generated together. It is estimated that the lack of a treatment solution for this waste stream costs from $30,000 to $60,000 annually in offsite disposal costs. Other sites such as the Marine Corps Logistics Base in Barstow, California have similar problems with surfactants. Therefore, methods to treat surfactants need to be developed.

1.2 Introduction

A surfactant is a substance that adsorbs onto surfaces or interfaces and alters the interfacial free energy of the surface. Also, a surfactant may act to chelate metals in solution, possibly causing difficulty with the treatment of metals in wastewater processes. Destruction of biodegradable surfactants typically occurs by biological treatment processes in wastewater treatment plants. However, biological treatment often is land-intensive, expensive, and requires careful maintenance. Surfactant use is based on a variety of properties including foamability, chemical stability, solubility, reactivity and corrosivity. For some applications it is not feasible to select a surfactant that is biodegradable, and it is possible that its toxicity properties are not well characterized or understood (Lewis, 1992).

Ultrasonic irradiation of aqueous solutions has been shown to be effective for the in situ destruction of a variety of organic and inorganic contaminants (Bhatnagar and Cheung, 1994; Drijvers et al., 1996; Hua and Hoffmann, 1996; Kontronarou et al., 1991; Kontronarou et al., 1992; Pétrier et al., 1992; Pétrier et al., 1994). Ultrasonic irradiation produces cavitation bubbles that yield internal bubble temperatures on the order of 5000 K and pressures of hundreds of atmospheres upon their implosion (Leighton, 1994). The interior of a cavitation bubble contains water vapor, gas and volatile compounds. Destruction of organic compounds occurs in the cavitation bubble itself or at the interfacial sheath by direct pyrolysis or hydroxylation that results from the gas-phase pyrolysis of H_2O:

$$H_2O \rightarrow H\bullet + OH\bullet \tag{1.2-1}$$

Reactions also are possible in the bulk aqueous phase by oxidation with hydrogen peroxide which is formed in sonolytic systems from the self reaction of $OH\bullet$, or the scavenging of $H\bullet$ by O_2 and the subsequent self reaction of $HO_2\bullet$:

$$OH\bullet + OH\bullet \rightarrow H_2O_2 \tag{1.2-2}$$

$$H\bullet + O_2 \rightarrow HO_2\bullet \tag{1.2-3}$$

$$HO_2\bullet + HO_2\bullet \rightarrow H_2O_2 + O_2 \tag{1.2-4}$$

In addition, polymers such as long-chained-polymeric surfactants are subject to destruction by pressure and shear forces generated as the bubble collapses. Sonication has been shown to be particularly effective with volatile and hydrophobic compounds since they can partition to the bubble or interfacial sheath rapidly. Surfactants have been shown to accumulate at the cavitation bubble interface (Alegria et al., 1989) and are expected to exhibit similar reactivities to volatile and hydrophobic compounds.

Sonoluminescence, a weak emission of light, also results from an acoustically cavitating bubble when high intensity ultrasound propagates through a liquid. Although sonoluminescence was first reported in 1934, the exact mechanism of the emission is still under debate. Lasting for less than 50 picoseconds, the light emission occurs in the final stages of bubble collapse when the contents of the bubble are under extreme temperature and pressure. A recent study has explored the role of surfactants and surface active agents on the intensity of sonoluminescence (Ashokkumar and Grieser, 1997). The study determined that the relative sonoluminescence intensity increased in the presence of an anionic surfactant, sodium dodecylsulfate (SDS), or a cationic surfactant, dodecyltrimethlyammonium chloride (DTAC). Sonochemistry and sonoluminescence have been shown to be interrelated since violently collapsing bubbles generating high temperatures and pressures are required in both cases. Thus, increased sonoluminescence is expected to correlate with increased sonochemical activity. Also, the presence of a surfactant at a bubble surface has been shown to allow more rapid growth of a cavitation bubble by a process called rectified diffusion (Fyrillas and Szeri, 1995). This is expected to result in a larger number of active cavitation bubbles and fewer bubbles that are growing to a radius where they will collapse violently. If bubbles are oscillating, but not cavitating, ultrasound is energetically inefficient. It is preferable to expend the ultrasonic energy to cause the cavitation bubbles to implode violently, and hence generate oxidizing free radicals.

The sound field, as defined by the frequency and pressure amplitude, affects how a cavitation bubble reacts under its influence. For an applied frequency, the bubble size at which maximum perturbations occur, resulting in the most chemical effects, is dependent on the ultrasonic frequency; smaller radii are influenced more by higher frequency waves. Similarly, the collapse time of a bubble is shorter at higher frequencies. The effect of frequency has been studied by many researchers (Hua and Hoffmann, 1997; Pétrier et al., 1996; Pétrier et al., 1992; Pétrier et al., 1994), with all of these studies reporting faster degradation at higher frequencies. The faster destruction at higher frequencies has been attributed to more hydroxyl radical formation due to shorter bubble growth times and more acoustic cycles per time than at low frequency. On the other hand, lower frequency cavitation is thought to be more violent, resulting in higher collapse temperatures in the bubble (Crum, 1995). Therefore, the more rapid growth of a cavitation bubble due to surfactants may result in increased sonochemical activity at low frequency compared to higher frequency by combining rapid bubble growth with higher temperatures of collapse.

Research in this report focuses on the use of sonication to destroy surfactants and surface tension properties in industrial wastewaters that affect traditional water treatment processes. We have performed a preliminary study to investigate the sonochemical destruction at 20 kHz of surfactants in wastewater from the Naval Underwater Warfare Center (NUWC) treatment plant. This study expanded that work significantly by investigating the destruction of various surfactants in three different reactors and understanding the release of metals from surfactants during sonication.

Although ethylenediaminetetraacetic acid (EDTA) is not a surfactant, it is a strong chelating agent. In addition to surfactant degradation studies, EDTA was chosen for initial metal complexation studies because it binds metals strongly but does not have surface-active properties. Therefore, metal-complex destruction is expected to be better with surface-active agents than with EDTA. Given the chelation ability of EDTA to metals, EDTA has undergone recent environmental scrutiny, also due to its widespread use, limited biodegradation and its potential to increase metal mobility. Thus, it is of interest to understand its degradation and fate in wastewater systems.

Finally, the effects of physical properties of surfactants and the effect of ultrasonic frequency were investigated to gain an understanding of factors affecting degradation. Successful partial or total destruction of surfactants resulting in the release of metals bound to surfactants may result in significant cost savings to treatment plants such as the wastewater treatment plant at NUWC.

CHAPTER 2.0

Experimental Methods

2.1 Materials

All chemicals were used as received. Water to prepare the solutions was from a MilliQ water purification system operating at R = 18.2 MΩ-cm.

An anionic surfactant, octylbenzene sulfonic acid (OBS), and a nonionic surfactant, t-octylphenoxy polyethoxyethanol (OPE), were chosen as model surfactants. 4-nitrophenol (PNP) and ethylbenzene sulfonate (EBS) were chosen as model non-surfactant compounds. Compounds were obtained from Sigma-Aldrich at purity >97% and were used without further purification. Table 2-1 shows physical-chemical properties of the surfactant compounds.

EDTA solutions were prepared using the disodium salt (approximately 99% pure, Sigma-Aldrich). An equimolar amount of chromium (III) nitrate (99.99%+, Sigma-Aldrich) or copper (II) nitrate (certified ACS, Fisher Scientific) was added to the EDTA solution to produce the metal complex. The chromium-EDTA complex was heated to approximately 70°C and then allowed to sit overnight while the copper-EDTA complex was allowed to sit for at least 24 hours at room temperature to ensure that complexation was complete.

Table 2-1. Properties of Surfactants and Non-surfactants.

	Structure	CMC (mM)	Γ (molec.cm^{-2} at 0.1 mM
	Surfactant		
OPE	$(CH_3)_3C(CH_2)C(CH_3)_2$ 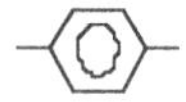 $O\text{-}[CH_2CH_2O]_n\text{-}H$ n=9-10	1.3	$1.6x10^{14}$
OBS	$CH_3(CH_2)_7$  SO_3	5.2	$1x10^{13}$
	Non surfactant		
EBS	$C_3\,H_5$ 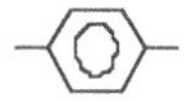 SO_3	--	--
PNP	OH  NO_2	--	--

2.2 Ultrasonic Units

Sonochemical experiments were conducted with three different ultrasonic reactors: a 20 kHz ultrasonic probe model 550 (Fisher Scientific), a near field acoustical processor (NAP) (Advanced Sonic Processing, NAP-3606-HP-TC) operating at 16 and 20 kHz and a URS 1000 Ultrasonic Laboratory Reactor consisting of a USW 51-52 Ultrasonic Transducer operating at 354 kHz with an LVG 60 Radio Frequency Generator (L3 Communications ELAC Nautik). For the probe system, samples were sonicated by immersing the probe tip into the solution to a depth of 4 cm in a 50-mL glass rosette reactor (Figure 2-1). Ultrasound was emitted from the replaceable titanium tip, which has an area of 1.20 cm^2. The NAP (Figure 2-2) has seven water-cooled magnetostrictive transducers fixed to each stainless steel parallel plate operating at 16 kHz on one plate and 20 kHz on the other with a sonoactive area of 1260 cm^2. The NAP system contains 2 L total solution volume with a 165 mL sonicating solution volume. It was run in a continuous circulation mode. A Teflon double diaphragm pump proved adequate for mixing in the temperature-controlled external glass vessel where samples were taken. In the ELAC system, 250-mL solutions were sonicated with the sound waves emitting from the bottom of the reactor (Figure 2-3). This exposes the solution to a sonoactive area of 23.4 cm^2. In both systems, the power supplied to the solution was measured using calorimetry (Mason and Lorimer, 1988). The reactors also contain a cooling water jacket, and a Fisher Scientific 1006S Isotemp cooling system was used to keep the reaction solution at a constant temperature of 25°C. The parameters of all systems as used in the experiments are listed in Table 2-2. Sonication experiments were performed under various gas-saturated conditions to alter the saturated gas characteristics: no-purge (initially air saturated), continuous air bubbling or continuous oxygen bubbling. Gas entered the solution through a glass micro-pipet inserted to the bottom of the ELAC reactor or with the horn in the probe system. The gas flow was not regulated, but was assumed to create consistent saturated gas conditions. Oxygen-bubbled systems were bubbled for 30 minutes prior to the start of sonication.

Figure 2-1. Fisher Scientific 20 kHz Ultrasonic Probe and Rosette Reactor.

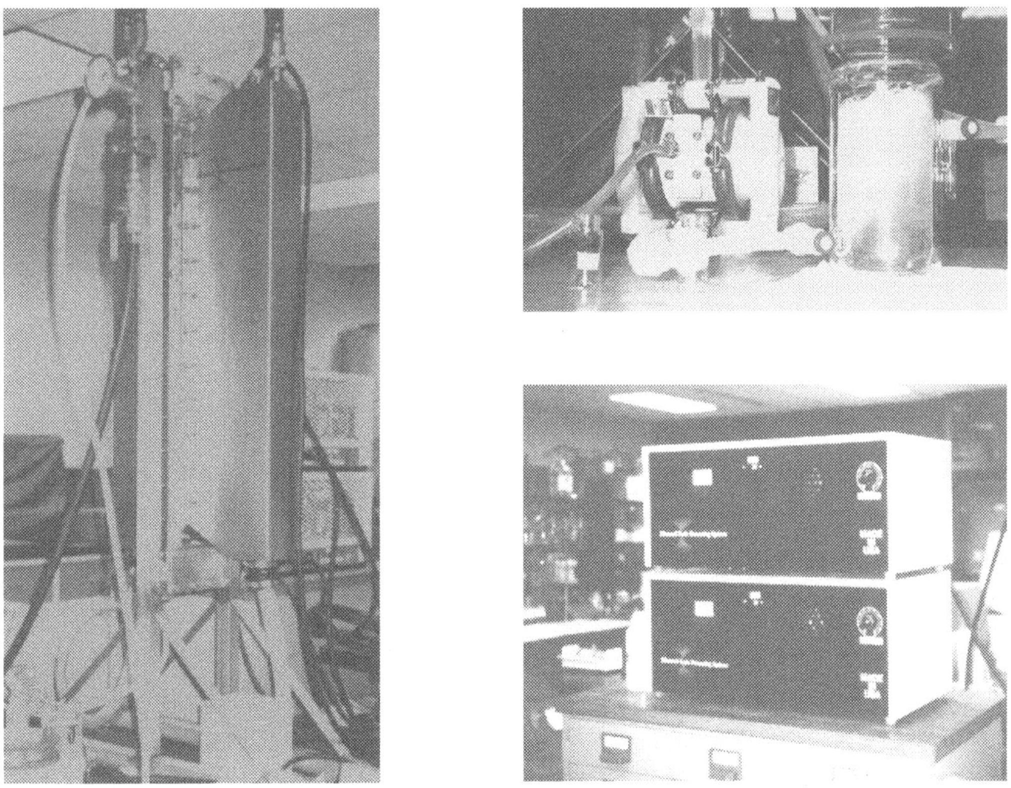

Figure 2-2. Near-Field Acoustical Processor (NAP) Reactor.
Left: transducers and sonochemical reactor
Top right: Teflon pump and external reactor
Bottom right: power sources

Figure 2-3. URS 1000 Ultrasonic Laboratory Reactor (Operated at 354 kHz).

Table 2-2. Operating Parameters for Ultrasonic Units.

	Sonication Volume (mL)	Power (W)	Power Density (W/mL)	Power Intensity (W/cm^2)
NAP (16 and 20 kHz)	165	107	0.053	0.085
Probe (20 kHz)	50	23.4	0.47	19.5
ELA Reactor (354 kHz)	250	37.3	0.15	1.6

2.3 Experimental Conditions

The solute concentration was 0.1 mM unless specified otherwise. A 33 mM phosphoric concentration at pH 2.8 was used as a buffer for all solutions except those containing EDTA. A pH of 2.8 was chosen for PNP and EBS because their pKa values are 7.1 and 7.0, respectively. All degradation experiments containing EDTA were run within a pH range of 6.5 to 7.5. No buffer was used to prevent additional complexes between the buffer species and the metal cations.

For EDTA experiments, the pH was monitored using an Orion 290A pH meter and pH adjustments were made by manual addition of 0.1 M sodium hydroxide solution (HPLC grade, Agilent Technologies) or 0.05 M $HClO_4$ solution (ACS grade, GFS Chemicals) in 20 μL aloquots using a 20 – 200 μL pipetter (Labnet). Additions did not exceed 1.2% of the original sonicated volume. Solutions were run with and without continuous gas bubbling. Compressed air and oxygen were high purity (Industrial grade, 99.9%, Praxair).

Samples (ranging in size from 0.5 to 2.0 mL) were withdrawn from the batch reactor at specified times during sonication using a Teflon syringe with Teflon needle for analysis. The total volume withdrawn during a single run never exceeded 10% of the total sonicated volume. Several batch reactions of various time lengths were run for experiments in which analyses required volumes of solution that exceeded this value. Samples withdrawn from the probe system were filtered using Millex-LCR 0.45 μm PTFE disposable filter units. Particles were not observed to form during sonication in the ELAC system; therefore, samples were not filtered. Duplicate degradation runs were performed to verify results at selected sonication times. Measured $[EDTA]_t$, $[Cu^{2+}/EDTA]_t$, and $[Cr^{3+}/EDTA]_t$ agreed to within 10% for samples from different runs. The error for calculated rate constants was calculated from the linear regression of the plotted points.

The relative reaction rate constants of the four compounds were measured using Fenton's method. A 25 mL mixture containing 0.5 mM of EBS and 0.5 mM of one of the compound were allowed to undergo OH radical reaction for 10 s. 2.5 mL of 1 N NaOH was added to stop the production of OH radical by reducing the Fe(II) to $Fe(OH)_3$. This procedure was repeated for 20 s, 40 s, 60 s and 70 s.

2.4 Analysis of Samples

Quantification of OPE, OBS and EBS was performed using a Hewlett Packard 1100 HPLC with a C-18 Hypersil ODS column. An eluent gradient of acetonitrile and phosphate buffer at pH 2.2 from 100% phosphate buffer to 55% phosphate buffer in 45 minutes was used for OBS and EBS detection. An isocratic eluent of 52% water and 48% acetonitrile was used for OPE. PNP was quantified using a Shimadzu Spectrophotometer UV-2401PC with the sample adjusted to pH>10 using 0.1 mL of 1 N NaOH per 0.5 mL sample. The surface tension of the solutions was determined using a Sensadyne Tensiometer (Chem Dyne Research Corporation, model PC 500).

Ion chromatography was performed using a Dionex model DX-100 ion chromatograph (IC) with an IonPac AS-5 anion exchange column and guard. The eluent was 1.6 mM sodium carbonate/0.72 mM sodium bicarbonate (both Fisher Scientific) and was run at a 2.00 mL min^{-1} flow rate. This method was used to quantify the degradation of EDTA and Cu^{2+}/EDTA, with an elution time of 1.6 and 3.6 minutes, respectively. A plot of the measured conductivity signal versus concentration of these compounds was linear within the studied range of 15 to 150 uM. Additional peaks were observed to form for sonicated solutions, but were not identified using IC.

A Shimadzu UV-2401PC UV-Vis spectrophotometer was used for Cr^{3+}/EDTA quantification. To achieve a measurable absorbance at 544 nm for the 15 to the 150 μM range, a 100-mm extended pathlength cell was used. For this analysis, each data point for a specific sonication time was obtained as a batch run due to the large quantity of solution required for analysis. The pH of measured samples and standards were all within the pH range of 7 to 7.5 to insure that observed absorbance changes were due to degradation of the complex and not the formation of different aqueous species of the Cr^{3+}-EDTA complex.

The production of hydrogen peroxide was measured using an iodometric technique (Kornmann et al., 1988). A 1.5 mL filtered sample of sonicated solution was mixed with 0.75 mL of 0.1 M potassium biphthalate in a 10-mm pathlength spectrophotometer cell. The absorbance at 352 nm was observed as a time plot with the addition of 0.75 mL of an iodide reagent (0.4 M potassium iodide, 0.06 M NaOH, and 0.0001 M ammonium molybdate). A maximum absorbance, or production of I_2, was reached after three minutes. A linear relationship was obtained for this maximum absorbance and the hydrogen peroxide concentration. Standards in the range of 15 to 150 μM were prepared by the serial dilution of a 30% stock solution (Mallinckrodt analytical reagent). The presence of organic peroxides can also be detected using this method. Organic peroxides react slower and can be identified as a slow increase in the solution absorbance following the rapid initial production of iodine. Sonication runs to determine peroxide production were conducted under conditions identical to the degradation experiments described before. Measurements of $[H_2O_2]$t for repeated sonication runs had a standard deviation of <7.5% of the average value.

The production of Cr(VI) in Cr^{3+}-EDTA sonicated solutions was verified using a UV-Vis spectrophotometer. Cr(VI) absorbs strongly at 425 nm, but quantification of this chromium species was not possible. Cr^{3+}/EDTA complex also absorbs at this frequency and produced a varying background signal with different sonication times.

Hydroxyl radical produced in the sonicated solutions was measured by direct OHΣ addition to terephthalate (TA) producing hydroxyterephthalic acid (HTA) (Hua and Hoffmann, 1997). Terephthalic acid (2.0 mM) was prepared in a 7.5 x 10^{-4} M NaOH and 4.5 x 10^{-3} M Na_2HPO_4 buffer (pH = 11). Continuously oxygen-bubbled TA solutions were sonicated at both 20 and 354 kHz. Aliquots (3.0 mL) were withdrawn at various times and placed into a quartz fluorometer 10-mm cell. Samples from the probe system were filtered as described before. Fluorescence measurements were made on a Shimadzu RF-5301PC spectrofluorometer with the following parameters: excitation and emission beam slit width = 1.5 nm, sampling interval = 1 nm, excitation wavelength = 315 nm, emission wavelength = 425 nm. Standard HTA solutions were prepared and a linear relationship for emission intensity (425 nm) and concentration was obtained for the range of concentrations produced during sonication experiments. Triplicate measurements for $[OH\bullet]_t$ agreed to within 5% of the average value.

In selected experiments to determine free and bound Cr^{3+}, a SP Sephadex C-25 cation exchange resin was used to remove any free Cr^{3+} released from the sonicated Cr-EDTA solutions. The exchange resin was prepared by soaking it for 24 hours in a 0.025 M Na_2HPO_4/0.025 M NaH_2PO_4 buffer solution (pH = 7). The swollen resin (0.3 g) was added to a 20-mL sonicated sample in a 25-mL polypropylene bottle and shaken for one hour. Supernatant was withdrawn and analyzed as is using a Varian Vista ICP-AES. The amount of Cr^{3+} taken up by the exchange resin was observed as a decrease in the total chromium concentration as measured by ICP-AES of the solution before and after the resin was added.

The total organic carbon (TOC) of the solutions for various sonication times was measured using a Shimadzu TOC-5000A in the non-purgeable organic carbon mode. Samples were filtered as described before when needed. They were then acidified with 2.0 M HCl to lower the solution pH below 3.0 and air-purged for three minutes before a triplicate sampling.

CHAPTER 3.0

RESULTS

3.1 Degradation of Model Surfactants at 20 kHz and Effect of Charge, Size, Substituents

For reaction with OH radicals, reaction rates are dependent on the concentration of the reactants at the interface and in the bulk. However, the predominant reaction will occur at the interface for reactants that are surface-active since OH radical is at a higher concentration at the interfacial region. Surface excess (σ) of a surface-active solute is the number of molecules that can accumulate at the interface and is calculated using

$$\Gamma = -\frac{1}{RT}\left(\frac{d\sigma}{d\ln C}\right) \tag{3.1-1}$$

for nonionic surfactants and for charged molecules,

$$\Gamma = -\frac{1}{RT}\left(\frac{d\sigma}{2d\ln C}\right) \tag{3.1-2}$$

where σ is the surface tension, C is the concentration of the surfactant, R is the universal gas constant and T is the temperature of the solution (Hiemenz and Rajagopalan, 1997). $\frac{d\sigma}{2d\ln C}$ can be calculated using the slope of the graph of σ vs. (ln C). There are three regions on a surface tension vs. ln concentration plot. In the first region at low concentration, surface excess increases with decreasing surface tension until the concentration transitions into the second region, the linear portion of the plot. At this point, the surface is saturated with surfactant molecules and any surfactant added will serve to increase the concentration in the bulk. In the third region, the linear region transitions into constant surface tension with concentration. Micelles, clusters of small molecules that form spontaneously in aqueous solutions, form in this region. The concentration at which the first micelle starts to form is known as the critical micelle concentration (CMC) (Hiemenz and

Rajagopalan, 1997). Above the CMC, surface excess remains the same as in the linear region. Figures 3-1 and 3-2 show the surface tension versus concentration plot of OPE and OBS, respectively. From the graph, the CMC of OPE was found to be 1.3 mM and that of OBS was found to be 5.2 mM at 20°C. Subsequently, the saturated surface excess was calculated to be 2.8 x 10^{14} molecules cm^{-2} and 1.4 x 10^{14} molecules cm^{-2} for OPE and OBS, respectively. Although during the collapse phase of the bubble, the temperature is on the order of thousands of Kelvin, this surface excess can still be used to quantify the system since the growth phase of the bubble is considered to be an isothermal process is at ambient conditions while accumulation of surfactants occurs on the bubble surface.

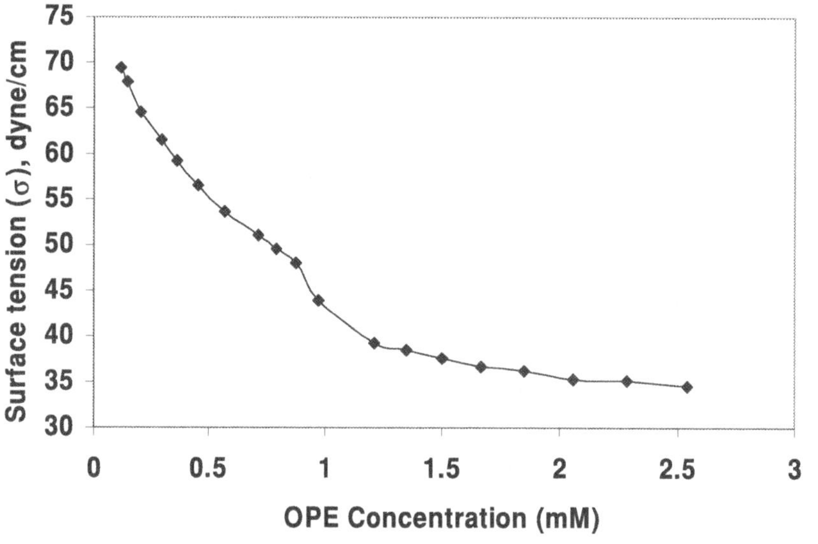

Figure 3-1. Surface Tension as a Function of Concentration for OPE at 30°C.

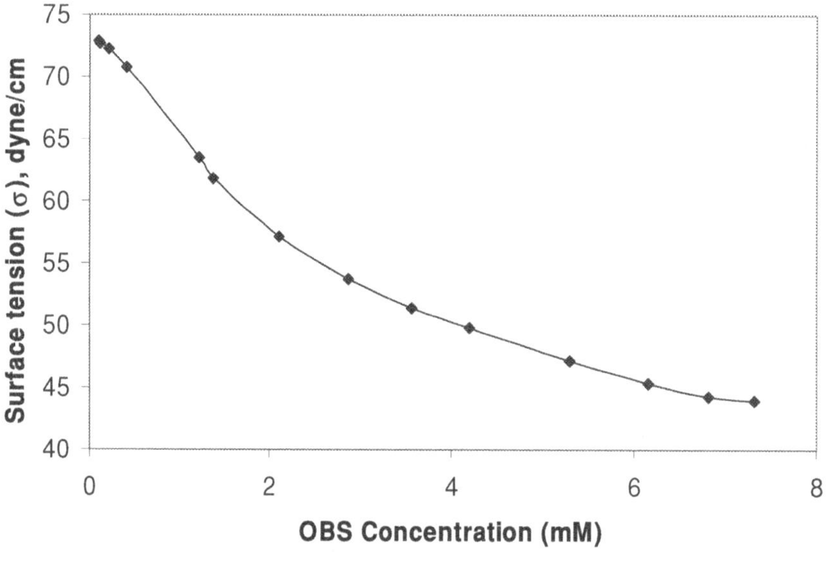

Figure 3-2. Surface Tension of OBS at 30°C.

Two 20 kHz reactors are used for low frequency studies. We have compared the degradation of model surfactants, octylphenoxypolyethoxyethanol (OPE) and 4-octylbenzenesulfonate (OBS), and non-surfactants, ethylbenzenesulfonate (EBS) and 4-nitrophenol (PNP), in both of these reactors as shown in Figures 3-3 and 3-4. In addition, EDTA was investigated in the 20 kHz probe. The anionic and nonionic surfactants were chosen to investigate the effect of a different substituent and charge on the degradation of an otherwise similar compound. EBS was chosen because it is identical to OBS except its hydrophobic tail, which is only two carbon atoms rather than the eight carbon atoms of OBS. PNP was investigated because it is widely studied in sonochemical systems making results easier to compare to other studies. Table 3-1 shows the first order degradation rate constants of t-octylphenoxy polyethoxyethanol (OPE), 4-octylbenzene sulfonate (OBS), ethylbenzene sulfonate (EBS) and 4-nitrophenol (PNP) in the probe and the NAP, respectively. Degradation rate constants were determined from the slope of lnC/Co versus time plots. It should be noted that degradation refers only to transformation of the parent surfactant compound, not complete destruction of all derivatives. Typically, longer sonication times are found to result in some decrease of the total organic carbon of the system indicating mineralization. Destaillats et al. (Destaillats et al., 2000) investigated the mechanism of OPE degradation by sonication. Interestingly, the major degradation mechanism led to degradation of the hydrophobic alkyl component of OPE rather than reaction on the hydrophilic ethoxylate component. The reaction on the ethoxylate component results in the formation of octylphenol, the more toxic degradation component observed in other degradation mechanisms. We expect that the degradation mechanism would be similar for other APEs with different chain length hydrophobic or hydrophilic components.

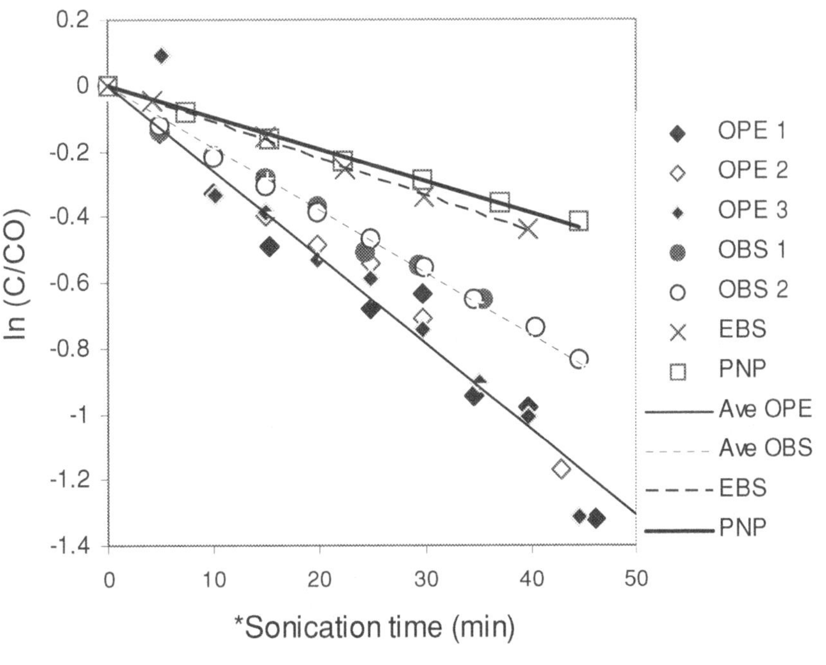

Figure 3-3. First-Order Plots of PNP, EBS, OBS and OPE Degradation Using NAP and 2L Total Volume. The solute in each solution was 0.1 mM, T=20C and dual ultrasound frequencies of 16 and 20 kHz. OBS, OPE, PNP and EBS were at pH 2.8. Multiple runs of OPE and OBS demonstrate the reproductibility of the results.

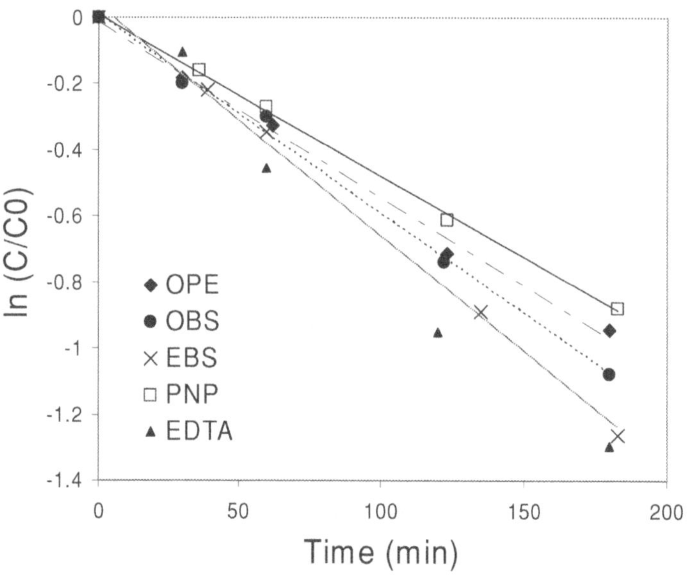

Figure 3-4. First-Order Plots of PNP, EBS, OBS, OPE and EDTA Degradation at 0.1 mM, T= and pH 2.8 in the 20 kHz Probe.

Table 3-1. Surface Excess, Diffusivity Ratios and Pseudo First-Order Rate Constraints of Degradation in the NAP and Probe Reactors at Consenstrations of 0.1 mM (0.15 mM for EDTA) and pH 2.8 (pH 7 for EDTA).

	A at 0.1mM molecule cm^{-2}	D_{iw}/D_{PNPw}	Fenton's reactivity $k_{i,fen}/k_{PNP,fen}$	$k_{NAP}x10^{-3}$ m in^{-1}	$k_{Probe}x10^{-3}$ m in^{-1}
OPE	$1.6x10^{14}$	0.37	0.9	26 ± 1	5.3 ± 0.2
OBS	$1.0x10^{13}$	0.60	2.8	19.0 ± 0.4	6.0 ± 0.2
EBS	---	0.85	1.2	11.2 ± 0.2	7.0 ± 0.2
PNP	---	1.0	1.0	9.7 ± 0.2	4.9 ± 0.1
EDTA	---	---	---	---	3.3

OPE, the nonionic surfactant, is found to have the fastest degradation rate in the NAP whereas EBS had the fastest degradation rate in the probe. As the surface tension decreases, the surface excess at the bubble interface increases as a result of more surfactant molecules gathering at the gas-liquid interface. This surface excess allows more OPE molecules to come into contact with the radicals existing at high concentrations at the liquid-air interface. In addition, a lower surface tension has been shown to result in increased bubble growth rates. Increased bubble growth rates allow more bubbles to reach a resonant size whereby they collapse violently.

The rate constants of OPE, OBS and EBS show a different trend from that in the NAP system although PNP is still the slowest reacting compound. This is attributed to the higher power intensity (18.2 W cm^{-2}) of the probe as compared to the NAP (0.0847 W cm^{-2}), resulting in a shorter lifetime for cavitation bubbles produced using the probe than the NAP. Although based on frequency, the properties of the cavitation bubbles (size, temperature of collapse, etc.) in both systems are expected to be the same, a shorter lifetime will limit the number of surfactant molecules that can reach the interface before collapse of the bubbles. Hence upon collapse, the cavitation bubbles in the NAP will have more surfactant molecules at the interface available for OH radical attack.

As the system is dynamic and rate constants and trends of degradation are different in the probe and NAP, non-equilibrium parameters need to be used to understand expected degradation rates. Typically, equilibrium parameters are used to determine expected trends in sonochemical degradation of compounds. Fick's first law applied to a system with compounds accumulating on an interface quantifies the flux (J_i, molecules/time/area of surface) of a compound from one phase to another and can be expressed as follows (Perry and Green, 1997):

$$J_i = -\frac{D_{iw}}{z}(C_{eq} - C_0) \tag{3.1-3}$$

where D_{iw} is the diffusion coefficient of the solute, i, in water which is inversely proportional to the molar volume of the solute, z is direction of the unidimensional diffusion (length), C_{eq} is the equivalent concentration of solute on the surface (*i.e.*, proportional to surface excess for surfactant, molar) and C_0 is the concentration of solute in bulk solution (molar). If C_{eq} is related to the surface excess (Γ) through a, the effective mass transfer interfacial area per unit volume (1/length), we derive the flux to be:

$$J_i = -\frac{D_{iw}}{z}(K_{eq}\Gamma a - C_0) \tag{3.1-4}$$

where

$$C_{eq} \propto \Gamma a \tag{3.1-5}$$

$$C_0 = C_{bulk} \tag{3.1-6}$$

and K_{eq} is the equilibrium relationship at the interface (unitless). For two solutions containing 0.1 mM of solute each, C_0 is assumed to be the same.

In the flux equation, two factors, namely the diffusivity (D_{iw}) and the equilibrium concentration at the interface (i.e., the surface excess), are important in determining the amount of a compound that reaches a cavitation bubble surface with time. The diffusion coefficients (D_{iw}) of the solutes studied, i, were unknown, but were estimated using the following equation (Perry and Green, 1997):

$$D_{iw} = \frac{8.621\times10^{-13}}{\mu_w^{1.14} V_i^{0.589}} \tag{3.1-7}$$

where μ_w is the solvent viscosity and Vi is the molar volume of the solute, i, at the solute normal boiling point (estimated based on molecular composition). The ratio of diffusivities compared to PNP, D_{iw}/D_{PNPw}, were calculated as 0.85, 0.603 and 0.368 for EBS, OBS and OPE, respectively. From the ratio, the diffusivity is the greatest for PNP followed by EBS, OBS and OPE. If none of the compounds preferentially accumulates on the surface, diffusivity is be the dominant term in Eqn 3.1-4. Thus, PNP is expected to have the fastest degradation rate constant since it has the faster ability to reach the interface. In the NAP system, the bubble has a longer lifetime due to the lower power intensity, hence equilibrium concentration plays a more dominant role in the flux and our results support this theory. For the probe system, the lifetime of the bubbles is shorter hence D_{iw} plays a more dominant role compared to the equilibrium concentration. Figure 3-5 shows the linear relation for OPE, OBS and EBS with diffusivity.

Equilibrium accumulation and diffusivity however, simply probe the location of the compounds in these systems. Other parameters of the compounds such as hydroxyl radical reaction rate constant are also important in determining rates of reactivity. Three possible explanations exist for PNP having the slowest reaction although it has the largest D_{iw}. First, its equilibrium surface excess concentration is expected to be much lower than the other three compounds hence the second part of the flux equation plays a bigger role in the flux to the bubble interface. The second possible explanation is that it is suspected to have a lower reactivity with hydroxyl radical than the other compounds. A third explanation may result from PNP having a different accumulation mechanism since it is not surface active but is hydrophobic. In order to determine if reactivity plays a role in the slower reaction rate constant of PNP, Fenton's reagent was used on mixtures of OPE, OBS or EBS and PNP to investigate their reaction rate constants with hydroxyl radicals in the absence of an air-water interface. Results indicate that PNP has a slower reaction rate constant with hydroxyl radical than either EBS or OBS as shown in Table 3-1. OBS has the fastest reaction rate constant followed by EBS. In addition, to determine if EBS or PNP may be accumulating on the bubble surface, we used a different surface tension measurement technique. It indicated that EBS has slight surface tension properties but PNP appears to be less surface-active. These results strengthen our interpretation of the results we have obtained.

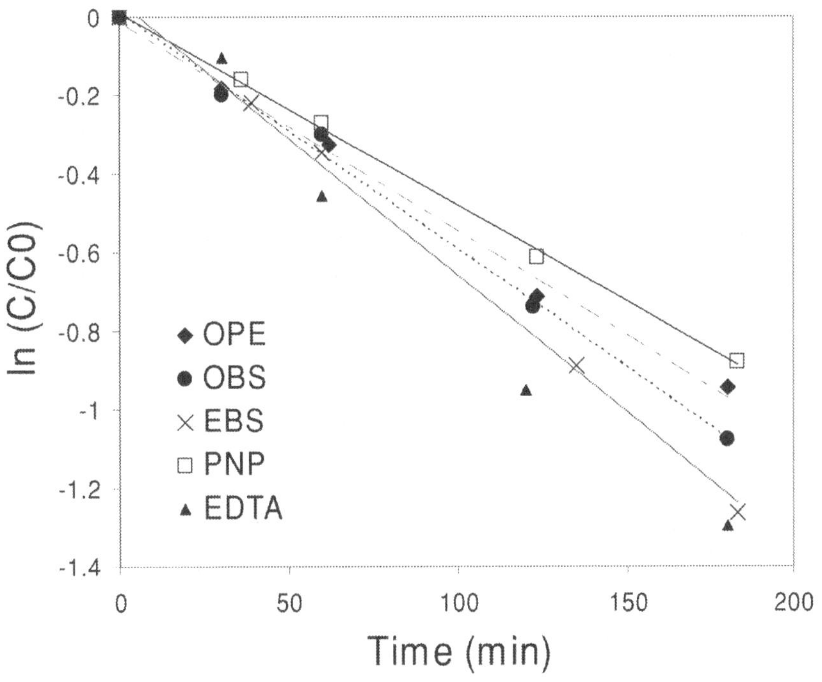

Figure 3-5. First-Order Rate Constant versus Diffusivities of EBS, OBS, OPE and PNP.

3.2 Degradation of Model Surfactants at 20 kHz, Effect of Matrix

3.2.1 pH Effects

The degradation of surfactants was performed at pH 7.4 to determine the effect of pH on degradation. As shown in Figures 2-6 and 2-7, pH did not have an effect on degradation of either OPE or OBS. The effect of pH on rates of reaction by sonolysis has been observed previously (Kontronarou et al., 1991; Kontronarou et al., 1992; Pétrier et al., 1994). Decreased degradation is observed as the compound is deprotonated due to its increased polarity in the deprotonated state. Since surfactants accumulate on surfaces independent of whether it is protonated or deprotonated, the effect of pH is minimal on degradation.

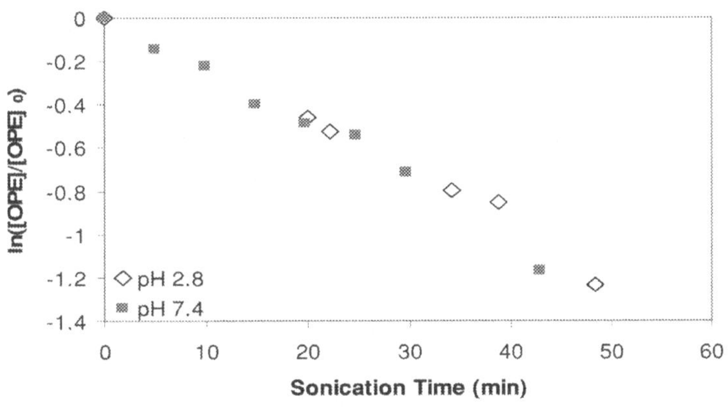

Figure 3-6. Comparison of Degradation of OPE at pH 2.8 and 7.4.

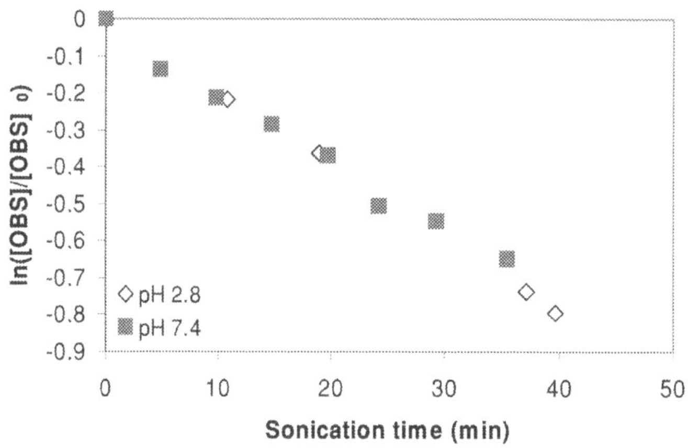

Figure 3-7. Comparison of Degradation of OBS at pH 2.8 and 7.4.

3.2.2 Concentration Effects

It has been observed in sonochemical systems that for increasing concentration of a contaminant, the pseudo first-order rate constant decreases (Kontronarou et al., 1991). The effect of concentration was investigated with surfactants to determine if ultrasound would be effective under a wide range of conditions possibly seen in an industrial wastewater treatment plant. As shown in Figure 3-8, the probe follows the trend observed with other compounds. However, in the NAP, at concentrations up to 0.1 mM, rate constants for OPE do not have the same trend as shown in Figure 3-9. We believe this is unique to surfactants because the longer lifetime of the cavitation bubbles in the NAP allows the surface tension reduction phenomena associated with surfactants to enhance the rate of reaction. Hence, there is a balancing effect between the concentration of the surfactant and the surface tension reduction observed for that concentration.

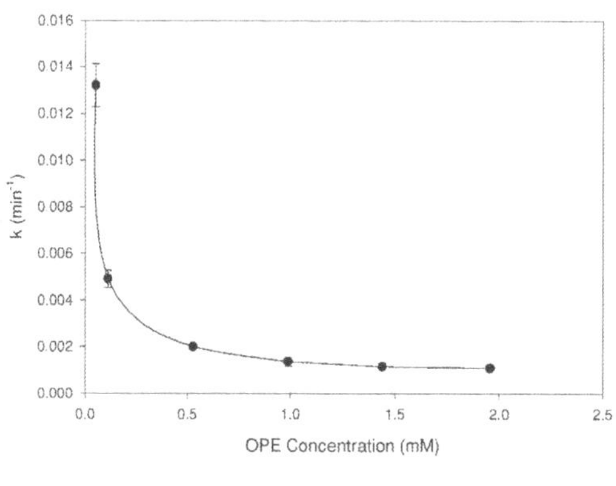

Figure 3-8. Probe First-Order Rate Constants as a Function of OPE Concentration. Temperature of Bulk Solution was 20°C.

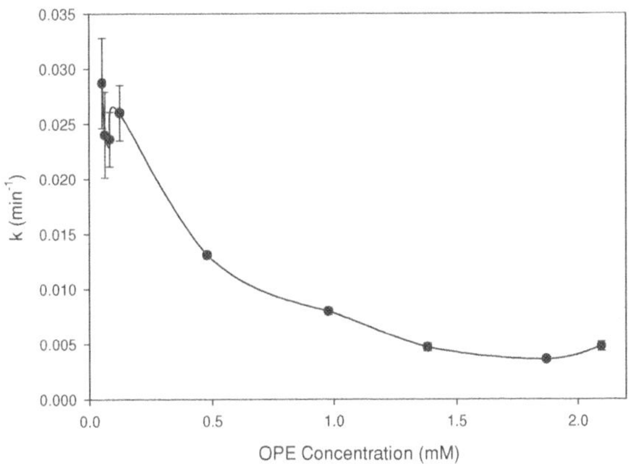

Figure 3-9. NAP First-Order Rate Constant as a Function of OPE Concentration. Bulk Solution Temperature was 20°C.

An initial rate represents the number of moles of reactant that undergo a reaction per time initially. It can be calculated using equation 3.2-1:

$$\frac{dC_0}{dt} = -kC_0 \tag{3.2.2-1}$$

where C_0 is the initial concentration and k is the observed pseudo first-order degradation rate constant. Figure 3-10 shows the variation in the initial rates with OPE and OBS in the NAP. The figures can be divided into three regions dominated by different factors. At low concentrations (0.05 mM to 0.5 mM for OPE, 0.05 mM to 1.5 mM for OBS), the amount of reactants that react per time (i.e., initial rates), increases almost linearly with concentration. This is attributed to a substantial increase in surface excess in this region. Therefore, more surfactant molecules move toward the interface for reaction. Also, there is a decrease in surface tension, which gives rise to a smaller Blake radius (the initial radius of a nucleus so that dissolution will not occur), increasing the number of nuclei able to grow to active cavitation bubbles.

In the second region (0.5 mM to 1.3 mM for OPE, 1.5 mM to 5.2 mM for OBS) there is a maximum reached and then a slight decrease in the rate. In this range of concentration, the surface excess is constant. However, there is a steep decrease in surface tension that will reduce the Blake radius substantially. The reduction in the Blake radius allows more nuclei to form into active cavitation bubbles. Thus, more surfaces are formed and flux toward the interface increases due to an increase in C_0. As the concentration at the interface increases, thermolysis becomes more predominant since more surfactant molecules are at the interface. This will affect OH radical production. Although thermolysis will also result in a reduction of surfactant molecules, it serves to inhibit OH radical production to a greater extent; hence, acting to reduce the reaction rate. In the first portion in this second region, the reduction in Blake radius is the dominating factor. In the second portion, initial rates decrease because thermolysis results in the reduction of OH radical production. Hence, although there was still an increase in the number of nuclei forming bubbles, the reaction via hydroxylation was reduced to a greater extent compared to the increase in the amount of surfactant undergoing reaction via thermolysis reactions.

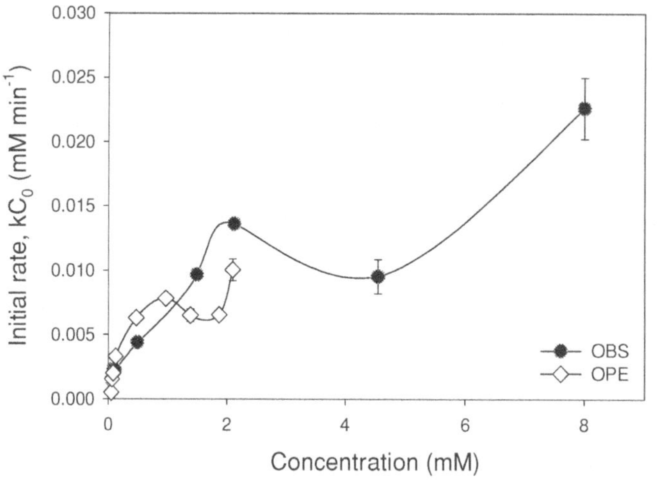

Figure 3-10. Initial Rates of OPE and OBS at Different Concentration in NAP.

The third region (>1.3 mM for OPE, >5.2 mM for OBS) shows a relatively constant initial rate with a large increase in the rate at high concentration. The constant portion is a result of constant surface tension and surface excess. There was an increase in initial rates at higher concentration due to the formation of micelles increasing the number of weak spots in solution where nucleation and subsequent growth of bubbles can occur. The presence of micelles may initiate more cavity formation, rather than sequestering surfactant molecules from the cavitation bubble interface. This effect is observed by the increase in the rates at 2.1 and 8 mM for OPE and OBS, respectively, in Figure 3-10. Although micelles were also present in the constant rate portion, micelle aggregates may be too small or the number of micelles may still be insufficient to affect the reaction rates. This demonstrates that this process is even more effective at higher concentrations than at low concentrations in the NAP reactor.

Rates of degradation of OPE in the probe are shown in Figure 3-11. The marked regions I, II and III designate the regions described above. It is clear that the effects observed with the NAP reactor are absent in the probe indicating significant differences in degradation between these two reactors. The linearly increasing trend in the probe is indicative of an adsorption limited competitive reaction observed in catalysis systems. This competitive reaction is likely a result of OH radical recombining in the interfacial region rather than reacting with target compounds.

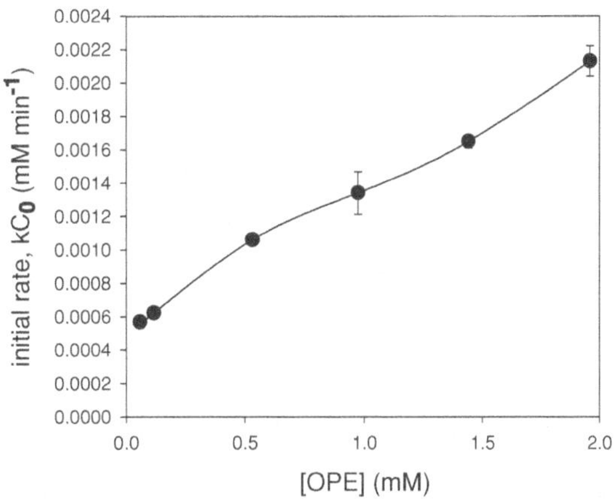

Figure 3-11. Initial Rates of OPE Degradation in the 20 kHz Probe Reactor.

3.2.3 Effect of Mixtures

In a typical industrial wastewater treatment plant, mixtures of surfactants as well as other organic and inorganic compounds are expected to be present. Therefore, we investigated the effect of simple mixtures on the degradation of surfactants. In a mixed system of 0.1 mM OPE and 0.1 mM OBS in the NAP, the rate constants are reduced by 36% for both compounds compared to individual systems. However, in a mixed system of 0.1 mM OPE and 0.1 mM PNP, the rate constant of OPE is approximately 77% of the single solute whereas the rate constant for PNP is reduced to 43% of the degradation in individual systems in the NAP. This indicates the selective degradation of OPE over PNP, a compound with less affinity for the surface. OPE and OBS accumulate on the interface to a similar degree, therefore, they directly compete for OH•. The reaction rate constant of OPE in the PNP/OPE mixture in the probe is approximately the same as that for OPE in the single solute solution. However, reaction rate constant of PNP in the mixture is approximately 37% of the PNP solution rate constant. Although these results are contrary to what we expected, they may be explained using the effect of flux as mentioned above. PNP is expected to have a lower flux due to less accumulation on the bubble surface hence PNP does not effectively compete with OPE molecules that reach closer to the cavitation bubble within the short lifetime of the bubble in the probe. PNP is expected to be farther away from the bubble surface in the probe than in the NAP due to less time for mass transfer. Thus, hydroxyl radical escaping reaction with OPE near the interface is not available to react with PNP before colliding with other radicals, other OH radicals, or OPE molecules in the bulk. For the NAP system, there is a longer lifetime for the OPE and PNP molecules to move near the interface. Hence, OPE and PNP will compete more directly for hydroxyl radical.

In both the probe and the NAP, degradation is based on the number of available sites for reaction; however, in the NAP power intensities allow for more contaminant to accumulate at the bubble surface resulting in increased degradation efficiencies. This is similar to our conclusions for results observed in the degradation of surfactants compared to non-surfactants in both reactors.

3.3 Degradation of Model Surfactants at 354 kHz

The frequency of sound affects the cavitation bubble size, time for bubble collapse and the velocity at which the bubble implodes. Higher frequency reactors are typically associated with higher OH radical generation whereas low frequency reactors are associated with higher temperatures and harsher conditions at collapse. Depending on the conditions desired, high or low frequency may be desired. Based on previous research using this equipment, we determined that 354 kHz would be the best frequency as this frequency has resulted in the fastest degradation rate of other compounds (Zhang and Hua, 2000). Figure 3-12 shows the degradation of all compounds in the 354 kHz reactor. Results show a differing trend from either of the 20 kHz reactors (NAP or probe). Degradation appears to be influenced by the reaction rate constant with OH radical and the diffusivity. This trend, although different from either low frequency reactor is expected since there is significantly more OH radical produced in the 354 kHz reactor (see section 4.7). Moreover, the higher frequency results in more growth and collapse cycles per time, thus reducing the time for compounds to accumulate on the bubble surfaces.

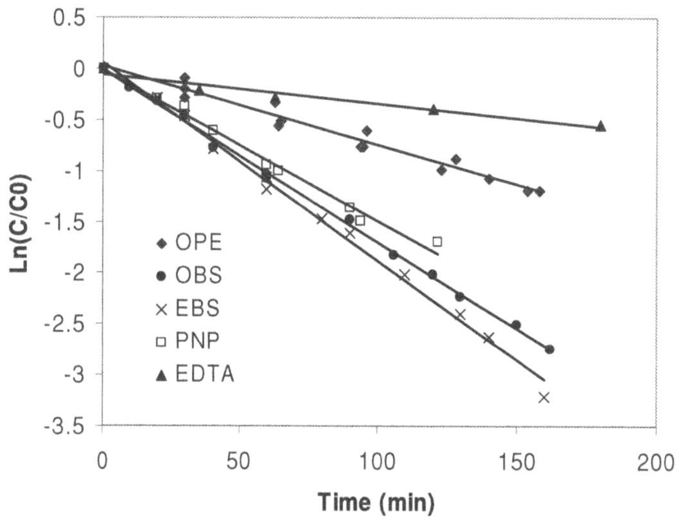

Figure 3-12. First-Order Plots of PNP, EBS, OBS, OPE and EDTA Degradation at 0.1 m M (0.15 mM for EDTA), T=20°C, and pH 2.8 (pH 7 for EDTA) in the 354 kHz Reactor.

3.4 Release of Metals from Surfactants at 20 kHz

Studies into chromium complexation focused on using ethylenediaminetetraacetic acid (EDTA) as the chelating species. EDTA was chosen for initial studies because it binds metals strongly but does not have surface-active properties. Therefore, metal-complex destruction is expected to be better with surface-active agents than with EDTA. Given the chelation ability of EDTA to metals, EDTA has undergone recent environmental scrutiny, also due to its widespread use, limited biodegradation and its potential to increase metal mobility. Thus, it is of interest to understand its degradation and fate in wastewater systems.

Only the 20 kHz probe reactor was used for low frequency studies with chromium. Control studies revealed the release of chromium into solution from the cavitational erosion of the stainless steel plates of the NAP reactor. The degradation of EDTA followed apparent pseudo first-order degradation as shown in Figure 3-4. The degradation rate constant was similar to that of the surfactant and non-surfactant compounds. EDTA degradation was also performed under constant oxygen bubbling. As shown in Figure 3-13, degradation is more rapid under oxygen than air. This effect of dissolved gases has been observed with other compounds (Hua et al., 1995; Hua and Hoffmann, 1997). The specific heat ratio, C_p/C_v or κ, and the thermal conductivity are similar for oxygen and nitrogen (Hua and Hoffmann, 1997; Mislik and Riesz, 1996). As these properties determine the bubble collapse temperature at a particular frequency (Hua et al., 1995; Leighton, 1994), it is reasonable to assume that the collapse conditions are essentially identical for both the O_2 and air-bubbled systems. Aqueous oxygen concentrations would remain higher in the oxygen-bubbled system, however. This would allow more oxygen to diffuse into the cavitation bubble and act as a hydrogen atom scavenger, reducing the likelihood of the recombination reaction of the hydrogen atom and hydroxyl radicals and favoring hydroxyl radical self-reaction. The production of hydroperoxyl radicals by oxygen scavenging of hydrogen atoms would also lead to higher hydrogen peroxide in O_2- saturated systems via the hydroperoxyl self-reaction.

Sonication of 150 uM Cr^{3+}-EDTA (1:1 ratio of metal to chelate) for two hours results in 20% reduction in the complex as shown in Figure 3-14. This reduction in Cr^{3+}-EDTA does not translate into an equal release in free-chromium as shown in Figure 3.15. Samples of complex solution sonicated for six hours were allowed to contact the resin for at least 30 minutes and show a 20% removal of total Cr^{3+} as measured by ICP-AES. The greater sonication time to achieve this amount of free chromium suggests that EDTA degradation products are still capable of binding the metal. Degradation of the chelate is not the only potential cause for metal release to be considered. Potential oxidation of Cr (III) to Cr (VI) would also affect the release of the metal from the chelate as it is transformed from a cation to dichromate ($Cr_2O_7^{2-}$). Initially, we used IC to detect chromate and dichromate and did not observe any formation.

An EDTA solution degrades to 30% degradation of a 150 uM EDTA solution (pH = 7) after two hours using the 20 kHz probe system with a 19.8 W/cm^2 power density. Degradation products observed in other OH radical producing systems (i.e., radiolysis and UV/TiO_2) were monitored in the sonolytic system. IC analysis of 20-kHz and 354-kHz sonicated EDTA, Cu^{2+}/EDTA, and Cr^{3+}/EDTA solutions did show the appearance of new peaks. The area of these peaks did increase with increased sonication time. Identification of these daughter compounds was attempted by performing identical IC analyses on standard solutions of selected compounds and comparing peak retention times. One peak was identified as either formic acid, acetic acid, or both as the two compounds are not separable with the IC method used. The area of the formic/acetic acid peak remained small over three hours of applied sound energy. All other peaks corresponding to degradation products remained unidentified. Maleic and malonic acid were eliminated as potential daughter compounds. IC analysis of standard solutions of ethylenediamine, nitrilotriacetic acid, glyoxylic acid, and iminodiacetic acid did not correspond to EDTA degradation peaks.

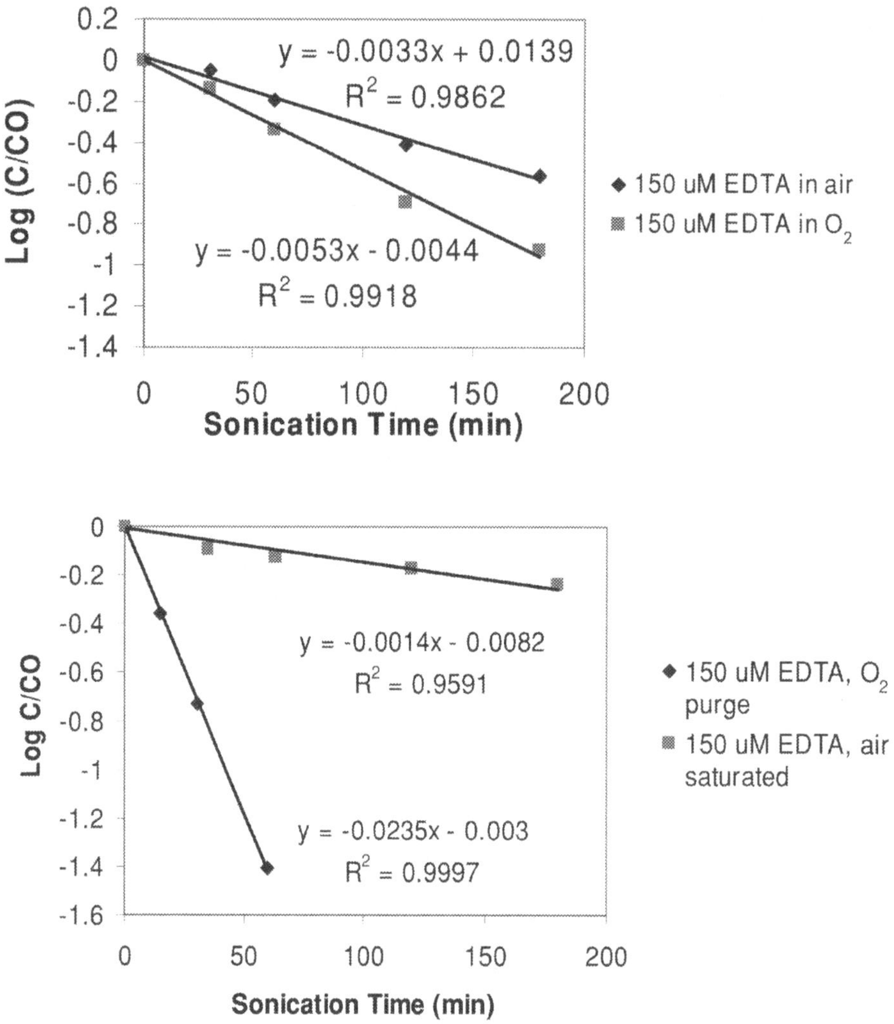

Figure 3-13. Comparison of EDTA Degradation with Air and Oxygen.
Top: degradation in 20 kHz probe
Bottom: degradation in 354 kHz reactor

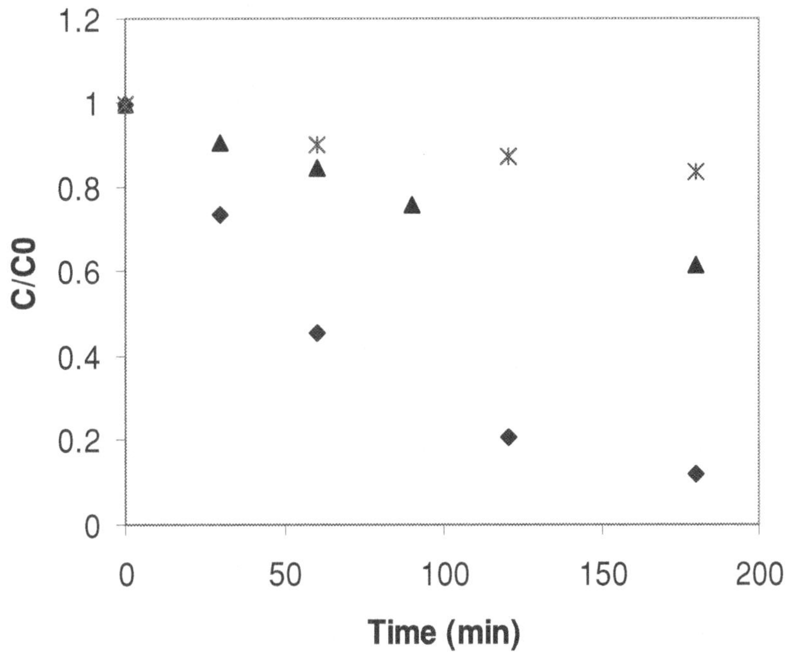

Figure 3-14. Comparison of the Sonochemical Degradation of 150 uM Solutions of Free and Metal-Complexed EDTA in the 20 kHz Probe with Continuous O_2 Bubbling and pH 7.

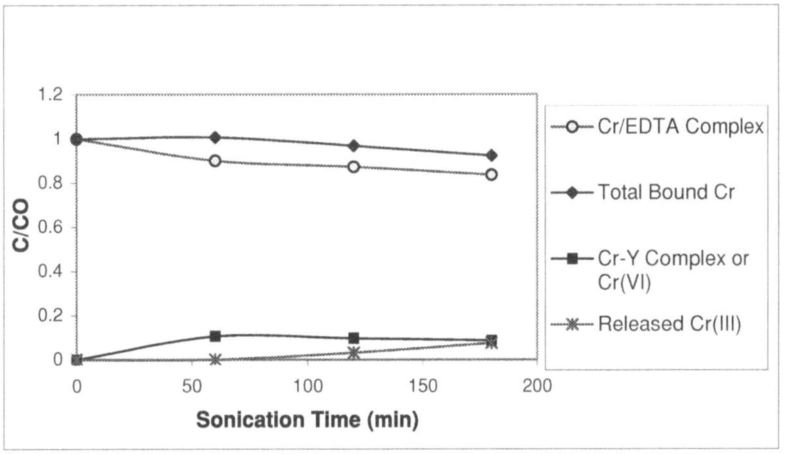

Figure 3-15. Sonication of 150 uM -EDTA in the 20 kHz Probe Reactor with Continuous Oxygen Purge. Temperature of Reaction was 20°C, pH≅7.

3.5 Release of Metals from Surfactants 354 kHz

Degradation of the Cr^{3+}-EDTA complex in the high frequency unit is shown in Figure 3-16. Compared to the low frequency reactor, degradation of the complex is similar. However, degraded complex resulted in either transformation of the complex or formation Cr(VI) as shown in Fig. 3-17. This may be due to fast oxidation to Cr(VI) as considerably higher amounts of OH radical are generated in this system as observed by H_2O_2 production rates.

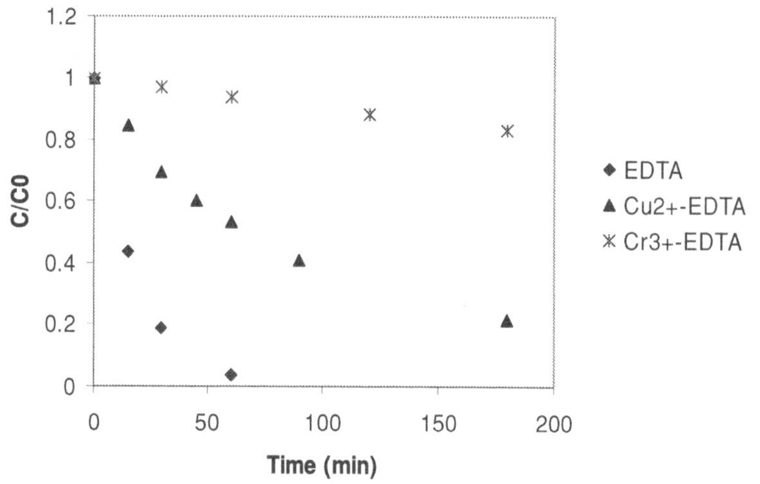

Figure 3-16. Comparison of the Sonochemical Degradation of 150 uM Solutions of Free and Metal-Complexed EDTA at 354 kHz with Continuous O_2 Bubbling and pH 7.

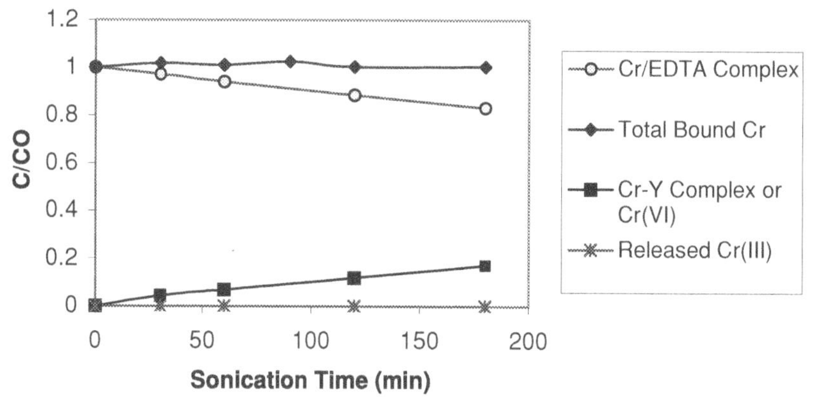

Figure 3-17. Sonication of 150 uM Cr^{3+}-EDTA in the 354 kHz Reactor with a Continuous Oxygen Purge. Temperature of Reaction was 20°C, pH≅7.

3.6 H_2O_2 Production and Degradation of Cu2+-EDTA Complexes at 20 kHz and 354 kHz

To gain insight into why the complex degrades slower than EDTA, we have investigated hydrogen peroxide production and the degradation of Cu^{2+}-EDTA complexes. Hydrogen peroxide has been used previously as an indication of the amount of OH radical in the system (Hua and Hoffmann, 1997). As shown in Figures 3-18 and 3-19, production of H_2O_2 is identical in water and the EDTA solution for an oxygen purged system. The similar production profile in EDTA and water indicates that EDTA reacts in bulk solution rather than at the cavitation bubble interface. By performing experiments with a continuous purge of air and oxygen in addition to similar systems without a continuous purge, we determined that the decreasing rate of production with time is due to degassing of the system by ultrasound. It is well-known that ultrasound can act to degas liquid media.

Degassing results in fewer cavitation nuclei available to grow to active cavitation bubbles, thus resulting in less radical formation. The oxygen-purged system reflects this with the linear production of hydrogen peroxide. Although EDTA and complexes react in the bulk solution rather than at the cavitation bubble interface in both reactors, hydrogen peroxide production is 10-fold higher in the 354 kHz reactor. The higher production rate of hydrogen peroxide indicates a higher production rate of OH radical as well.

Copper is known to have faster exchange rates with ligands than chromium. Thus, to test if the slower degradation of Cr^{3+}-EDTA complexes was due to chromium's slow exchange kinetics with ligands, we compared it to the degradation of Cu^{2+}-EDTA complexes. As expected, the degradation of the Cu^{2+}-EDTA complexes was considerably faster than Cr^{3+}-EDTA as shown in Figures 2-14 and 2-16 but slower than EDTA without metals complexed. We do not anticipate this slow degradation in the complexed systems to be as extreme (if apparent at all) in metal-surfactant systems. EDTA is a small molecule with six binding sites to the metals, thus, attack by OH radical occurs when the ligand releases from one of the six binding sites. With a larger molecule all of the reaction sites will not be occupied by the metal, thus reactions are not expected to be dependent on ligand exchange kinetics.

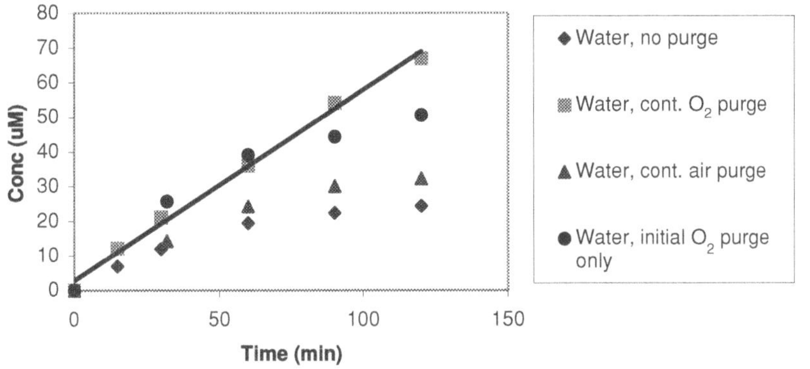

Figure 3-18. H_2O_2 Formation in the 20 kHz Probe Under Various Conditions.

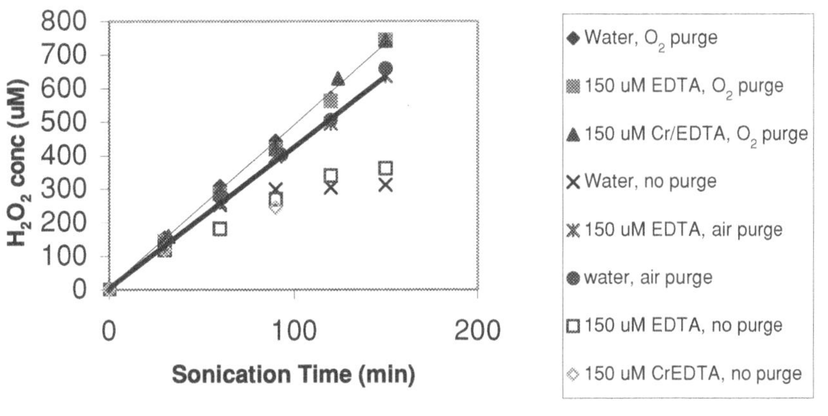

Figure 3-19. H_2O_2 Formation in the 354 kHz Probe Under Various Conditions.

CHAPTER 4.0

CONCLUSIONS

Sonochemical degradation of surfactants was observed to be more effective than non-surfactant compounds. In addition, as the concentration is increased the degradation rate constant does not decrease as significantly as with non-surfactant compounds in the NAP reactor. In fact, the total number of molecules degraded increases with concentration. The degradation of metal complexes is not as effective as in the absence of the metal. However, this is likely an artifact of the model complexing agent used. Surfactant metal complexes are expected to be faster as they will accumulate at the hot bubble interface, significantly increasing ligand exchange kinetics and thus degradation of the complex.

APPENDIX A

DISSEMINATION OF RESULTS

Gim-Yang Pee, the student working with OPE and OBS, defended her Master of Science thesis in November, 2000. One paper on her work was submitted to *Water Environment Research* and another is being prepared. J. Aaron Frim defended his M.S. thesis in August, 2001. One manuscript has been submitted to *Water Research* based on his work. Both students presented their results to the Division of Industrial and Engineering Chemistry at the 222nd National Meeting of the American Chemical Society meeting in Chicago in August, 2001. In addition, the PI spoke at the *Chemical Transformations and Phase Transfer Symposium* in the Division of Environmental Chemistry at the 221st National Meeting of the American Chemical Society in April, 2001. Finally, the PI presented this work at WEFTEC. Following is a list of publications and presentations.

G. Y. Pee, L. K. Weavers, and J. F. Rathman, 220th National Meeting of the American Chemical Society, Division of Environmental Chemistry, August 20-25, 2000, "Sonochemical Destruction of Surfactants."

G.Y. Pee, L. K. Weavers, and J. F. Rathman, The 23rd Midwest Environmental Chemistry Workshop, East Lansing, MI, October 6-8, 2000, "Sonochemical Degradation of Surfactants."

J. A. Frim and L. K. Weavers, The 23rd Midwest Environmental Chemistry Workshop, East Lansing, MI, October 6-8, 2000, "Sonochemical Destruction of Chromium Complexes."

G. Y. Pee, *The Application of Sonochemistry for the Destruction of Surface-Active Contaminants.* M.S. Thesis, Department of Chemical Engineering, The Ohio State University, Columbus, OH, 2000.

L. K. Weavers, G. Y. Pee, and J. F. Rathman, 221st National Meeting of the American Chemical Society, Division of Environmental Chemistry, San Diego, CA, April 1-5, 2001, "Kinetics of Surfactant Degradation by Sonolysis."

J. Aaron Frim, *The Sonochemical Release of Chelated Metals*. M.S. Thesis, Department of Civil and Environmental Engineering and Geodetic Science and Department of Geological Sciences, The Ohio State University, Columbus, OH, 2001.

J. A. Frim and L. K. Weavers, 222nd National Meeting of the American Chemical Society, Division of Industrial and Engineering Chemistry, Chicago, IL, August 26-30, 2001, "Sonochemical Release of Chelated Metals."

G. Y. Pee, L. K. Weavers, and J. F. Rathman, 222nd National Meeting of the American Chemical Society, Division of Industrial and Engineering Chemistry, Chicago, IL, August 26-30, 2001, "Enhanced Degradation of Surface-active Compounds by Sonolysis."

J. A. Frim and L. K. Weavers. "Sonochemical Destruction of Free and Metal-Binding Ethylenediaminetetraacetic Acid," submitted to *Water Research.*

L. K. Weavers, G. Y. Pee, J. A. Frim, L. Yang, and J. F. Rathman. "Ultrasonic Destruction of Surfactants: Application to Industrial Wastewaters," submitted to *Water Environment Research.*

L. K. Weavers and J. A. Frim, Eighth Conference of the European Society of Sonochemistry, Villasimius, Italy, September 14-19, 2002, "Sonochemical Destruction of Free and Metal-Binding Ethylenediaminetetraacetic Acid."

L. K. Weavers, J. A. Frim, G. Y. Pee, and J. F. Rathman. "Ultrasonic Destruction of Surfactants: Application to Industrial Wastewaters." *Proceedings of WEFTEC 2002*, Chicago, IL, September 28-October 2, 2002.

References

Alegria, A., Y. Lion, T. Kondo and P. Riesz. 1989. Sonolysis of Aqueous Surfactant Solutions. Probing the Interfacial Region of Cavitation Bubbles by Spin Trapping. *J. Phys. Chem.* 93, 4908-4913.

Ashokkumar, M., and F. Grieser. 1997. Sonoluminescence from Aqueous Alcohol and Surfactant Solutions. *NATO Advanced Study Institute on Sonochemistry and Sonoluminescence.*

Bhatnagar, A., and H. M. Cheung. 1994. Sonochemical Destruction of Chlorinated C1 and C2 Volatile Organic Compounds in Dilute Aqueous Solution. *Environ. Sci. Technol.* 28, 1481-1486.

Crum, L. A. 1995. Comments on the Evolving Field of Sonochemistry by a Cavitation Physicist. *Ultrasonics Sonochemistry* 2, S147-152.

Destaillats, H., H. M. Hung, and M. R. Hoffmann. 2000. Degradation of Alkylphenol Ethoxylate Surfactants in Water with Ultrasonic Irradiation. *Environ. Sci. Technol.* 34, 311-317.

Drijvers, D., R. De Baets, A. De Visscher, and H. Van Langenhove. 1996. Sonolysis of Trichloroethylene in Aqueous Solution: Volatile Organic Intermediates. *Ultrasonics Sonochemistry* 3, S83-S90.

Fyrillas, M. M., and A. J. Szeri. 1995. Dissolution or Growth of Soluble Spherical Oscillating Bubbles: The Effect of Surfactants. *J. Fluid Mech.* 289, 295-314.

Hiemenz, P. C., and R. Rajagopalan. 1997. *Principles of Colloid and Surface Chemistry.* Marcel Dekker.

Hua, I., R. H. Höchemer, and M. R. Hoffmann. 1995. Sonolytic Hydrolysis of p-Nitrophenyl Acetate: The Role of Supercritical Water. *J. Phys. Chem.* 99, 2335-2342.

Hua, I., and M. R. Hoffmann. 1996. Kinetics and Mechanism of the Sonolytic Degradation of CCl4: Intermediates and By-Products. *Environ. Sci. Technol.* 30, 864-871.

Hua, I., and M. R. Hoffmann. 1997. Optimization of Ultrasonic Irradiation as an Advanced Oxidation Technology. *Environ. Sci. Technol.* 31, 2237-2243.

Kontronarou, A., G. Mills and M. R. Hoffmann. 1991. Ultrasonic Irradiation of p-Nitrophenol in Aqueous Solution. *J. Phys. Chem.* 95, 3630-3638.

Kontronarou, A., G. Mills, and M. R. Hoffmann. 1992. Oxidation of Hydrogen Sulfide in Aqueous Solution by Ultrasonic Irradiation. *Environ. Sci. Technol.* 26, 2420-2428.

Kornmann, C., D. Bahnemann, and M. R. Hoffmann. 1988. Photocatalytic production of H2O2 and organic peroxides in aqueous suspensions of TiO2, ZnO, and desert sand. *Environ. Sci. Technol.* 22, 798-806.

Leighton, T. G. 1994. *The Acoustic Bubble*. Academic Press.

Lewis, M. A. 1992. The Effects of Mixtures and Other Environmental Modifying Factors on the Toxicities of Surfactants to Freshwater and Marine Life. *Wat. Res.* 26, 1013-1023.

Mason, T. J., and J. P. Lorimer. 1988. *Sonochemistry: Theory, Applications and Uses of Ultrasound in Chemistry*. Ellis Horwood Ltd.

Mislik, V., and P. Riesz. 1996. Nitric Oxide Formation by Ultrasound in Aqueous Solution. *J. Phys. Chem.* 100, 17986-17994.

Perry, R. H., and D. W. Green. 1997. *Perry's Chemical Engineers' Handbook*. McGraw-Hill.

Pétrier, C., B. David, and S. Laguian. 1996. Ultrasonic Degradation at 20 kHz and 500 kHz of Atrazine and Pentachlorophenol in Aqueous Solution: Preliminary Results. *Chemosphere* 32, 1709-1718.

Pétrier, C., A. Jeunet, J.-L. Luche, and G. Reverdy. 1992. Unexpected Frequency Effects on the Rate of Oxidative Processes Induced by Ultrasound. *J. Am. Chem. Soc.* 114, 3148-3150.

Pétrier, C., M.-F. Lamy, A. Francony, A. Benahcene, and B. David. 1994. Sonochemical Degradation of Phenol in Dilute Aqueous Solutions: Comparison of the Reaction Rates at 20 and 487 kHz. *J. Phys. Chem.* 98, 10514-10520.

Renner, R. 1997. European Bans on Surfactant Trigger Transatlantic Debate. *Environ. Sci. Technol.* 31, 316A-320A.

Zhang, G., and I. Hua. 2000. Cavitation Chemistry of Polychlorinated Biphenyls: Decomposition Mechanisms and Rates. *Environ. Sci. Technol.* 34, 1529-1534.

WASTEWATER UTILITY

Alabama
Montgomery Water Works & Sanitary Sewer Board

Alaska
Anchorage Water & Wastewater Utility

Arizona
Gila Resources
Glendale, City of, Utilities Department
Mesa, City of
Phoenix Water Services Department
Pima County Wastewater Management

Arkansas
Little Rock Wastewater Utility

California
Calaveras County Water District
Central Contra Costa Sanitary District
Contra Costa Water District
Crestline Sanitation District
Delta Diablo Sanitation District
Dublin San Ramon Services District
East Bay Municipal Utility District
El Dorado Irrigation District
Escondido, City of
Fairfield-Suisun Sewer District
Irvine Ranch Water District
Las Virgenes Municipal Water District
Lodi, City of
Los Angeles, City of
Los Angeles County, Sanitation Districts of
Napa Sanitation District
Orange County Sanitation District
Palo Alto, City of
Riverside, City of
Sacramento Regional County Sanitation District
San Diego Metropolitan Wastewater Department, City of
San Francisco, City & County of
San Jose, City of
Santa Barbara, City of
Santa Rosa, City of
South Bayside System Authority
South Orange County Wastewater Authority
Union Sanitary District
West Valley Sanitation District

Colorado
Boulder, City of
Colorado Springs, City of
Littleton/Englewood Water Pollution Control Plant
Metro Wastewater Reclamation District, Denver

Connecticut
New Haven, City of, WPCA

District of Columbia
District of Columbia Water & Sewer Authority

Florida
Broward, County of
Fort Lauderdale, City of
Gainesville Regional Utilities
JEA
Kissimmee, City of, Department of Water Resources
Miami-Dade Water & Sewer Authority
Orange County Utilities Department
Orlando, City of
Reedy Creek Improvement District
Seminole County Environmental Services
St. Petersburg, City of
Stuart Public Utilities
Tallahassee, City of
Tampa, City of
West Palm Beach, City of

Georgia
Atlanta Department of Watershed Management
Augusta, City of
Clayton County Water Authority
Cobb County Water System
Columbus Water Works
Fulton County
Gwinnett County Department of Public Utilities
Macon Water Authority
Savannah, City of

Hawaii
Honolulu, City and County of

Idaho
Pocatello, City of

Illinois
American Bottoms Wastewater Treatment Plant
Dupage, County of
Greater Peoria Sanitary District
Kankakee River Metropolitan Agency
Metropolitan Water Reclamation District of Greater Chicago
North Shore Sanitary District
Wheaton Sanitary District

Indiana
Fort Wayne, City of

Iowa
Ames, City of
Des Moines Metro Wastewater Reclamation Authority

Kansas
Johnson County Unified Wastewater Districts
Unified Government of Wyandotte County/Kansas City, City of

Kentucky
Louisville & Jefferson County Metropolitan Sewer District

Louisiana
Sewerage & Water Board of New Orleans

Maine
Bangor, City of
Portland Water District

Maryland
Anne Arundel County Bureau of Utility Operations
Howard County Department of Public Works
Washington Suburban Sanitary Commission

Massachusetts
Boston Water & Sewer Commission
Upper Blackstone Water Pollution Abatement District

Michigan
Ann Arbor, City of
Detroit, City of
Grand Rapids, City of
Holland Board of Public Works
Lansing, City of
Owosso Mid-Shiawassee County WWTP
Saginaw, City of
Wayne County Department of Environment
Wyoming, City of

Minnesota
Rochester, City of

Missouri
Independence, City of
Kansas City Missouri Water Services Department
Little Blue Valley Sewer District
Metropolitan St. Louis Sewer District

Nebraska
Lincoln Wastewater System

Nevada
Henderson, City of

New Jersey
Passaic Valley Sewerage Commissioners

New York
New York City Department of Environmental Protection

North Carolina
Cary, Town of
Charlotte/Mecklenburg Utilities
Durham, City of
Metropolitan Sewerage District of Buncombe County
Orange Water & Sewer Authority

Ohio
Akron, City of
Butler County Department of Environmental Services
Columbus, City of
Metropolitan Sewer District of Greater Cincinnati
Northeast Ohio Regional Sewer District
Summit, County of

Oklahoma
Tulsa, City of

Oregon
Clean Water Services
Eugene/Springfield Water Pollution Control
Water Environment Services

Pennsylvania
Philadelphia, City of
University Area Joint Authority, State College

South Carolina
Charleston Commissioners of Public Works
Mount Pleasant Waterworks & Sewer Commission
Spartanburg Sanitary Sewer District

WERF SUBSCRIBERS

Tennessee
Cleveland, City of
Knoxville Utilities Board
Murfreesboro Water & Sewer Department
Nashville Metro Water Services

Texas
Amarillo, City of
Austin, City of
Dallas Water Utilities
Denton, City of
El Paso Water Utilities
Fort Worth, City of
Gulf Coast Waste Disposal Authority
Houston, City of
San Antonio Water System
Trinity River Authority

Utah
Salt Lake City Corporation

Virginia
Alexandria Sanitation Authority
Arlington, County of
Fairfax County Virginia
Hampton Roads Sanitation District
Henrico, County of
Hopewell Regional Wastewater Treatment Facility
Loudoun County Sanitation Authority
Lynchburg Regional WWTP
Prince William County Service Authority
Richmond, City of

Washington
Edmonds, City of
Everett, City of
King County Department of Natural Resources
Seattle Public Utilities
Yakima, City of

Wisconsin
Green Bay Metro Sewerage District
Madison Metropolitan Sewerage District
Milwaukee Metropolitan Sewerage District
Racine, City of
Sheboygan Regional Wastewater Treatment
Wausau Water Works

Australia
South Australian Water Corp.
Sydney Water Corp.
Water Corp. of Western Australia

Canada
Greater Vancouver Regional District
Toronto, City of, Ontario

Mexico
Servicios de Agua y Drenaje de Monterrey, I.P.D.

United Kingdom
Yorkshire Water Services Limited

STORMWATER UTLITY

California
Los Angeles, City of, Department of Public Works
Monterey, City of
Santa Rosa, City of

Colorado
Boulder, City of

Georgia
Griffin, City of

Kansas
Overland Park, City of

Kentucky
Louisville & Jefferson County Metropolitan Sewer District

Maine
Portland Water District

North Carolina
Cary, Town of
Charlotte, City of, Stormwater Services

Pennsylvania
Philadelphia, City of

STATE

Fresno Metropolitan Flood Control District, Calif.
Urban Drainage & Flood Control District, Colo.

CORPORATE

ADS Environmental Services
The ADVENT Group Inc.
Alan Plummer & Associates
Alden Research Laboratory
Alpine Technology Inc.
Aquateam–Norwegian Water Technology Centre A/S
BaySaver Inc.
BioVir Inc.
Black & Veatch
Boyle Engineering Corporation
BPR CSO
Brown & Caldwell
Burns & McDonnell
CABE Associates Inc.
The Cadmus Group
Camp Dresser & McKee Inc.
Carollo Engineers Inc.
Carpenter Environmental Associates Inc.
CDS Technologies Inc.
Chemtrac Systems Inc.
CH2M HILL
Clancy Environmental Consultants Inc.
Damon S. Williams Associates, LLC
David L. Sheridan, P.C.
Earth Tech Inc.
Ecolab Water Care Services
EMA Inc.
The Eshelman Company Inc.
Finkbeiner, Pettis, & Strout Inc.
Frontier Geosciences Inc.
ftn Associates Inc.
Gannett Fleming Inc.
GE Betz
Golder Associates Inc.
Greeley and Hansen LLC
The HACH Company
Hazen & Sawyer, P.C.
HDR Engineering Inc.
HNTB Corporation
HydroQual Inc.
Insituform Technologies Inc.
Institute for Environmental Technology & Industry, Korea
Jacobson Helgoth Consultants Inc.
Jason Consultants Inc.
Jordan, Jones, & Goulding Inc.
KCI Technologies Inc.
Kelly & Weaver, P.C.
Kennedy/Jenks Consultants
Komline Sanderson Engineering Corporation
Lawler, Matusky & Skelly Engineers, LLP
Limno-Tech Inc.
Lombardo Associates Inc.
Malcolm Pirnie Inc.
McKim & Creed
MEC Analytical Systems Inc.
MWH
New England Organics
Odor & Corrosion Technology Consultants Inc. (OCTC)
ONDEO Degremont Inc.
Oswald Green, LLC
PA Government Services Inc.
Parametrix Inc.
Parsons
Post, Buckley, Schuh & Jernigan
R&D Engineering/Conestoga Rover & Associates
The RETEC Group
R.M. Towill Corporation
Ross & Associates Ltd.
Royce Technologies
SAIC Maritime Technical Group
Sear-Brown
Sierra Environmental Services
SYNAGRO
Tetra Tech Inc.
Trojan Technologies Inc.
URS Corporation
USFilter NATC
Vortechnics Inc.
Wade-Trim Inc.
Weston Solutions Inc.
Woodard & Curran
WRc/D&B, LLC
WWETCO, LLC

INDUSTRY

American Electric Power
ChevronTexaco Energy Research & Technology Company
The Coca-Cola Company
Dow Chemical Company
DuPont Company
Eastman Kodak Company
Merck & Company Inc.
ONDEO Services
Procter & Gamble Company
PSEG Services Corp.
Severn Trent Services Inc.
Shell Global Solutions
Thames Water Plc
United Water Services LLC

Note: List as of 5/12/03

Board of Directors

Research Council

Stormwater Technical Advisory Committee

WERF Product Order Form

As a benefit of joining the Water Environment Research Foundation, subscribers are entitled to receive one complimentary copy of all final reports and other products. Additional copies are available at cost (usually $10). To order your complimentary copy of a report, please write "free" in the unit price column. WERF keeps track of all orders. If the charge differs from what is shown here, we will call to confirm the total before processing.

Name ____________________ Title ____________________

Organization ____________________

Address ____________________

City __________ State __________ Zip Code __________ Country __________

Phone __________ Fax __________ Email __________

Stock #	Product	Quantity	Unit Price	Total
			Postage & Handling	
			VA Residents Add 4.5% Sales Tax	
			Canadian Residents Add 7% GST	
			TOTAL	

Method of Payment: (All orders must be prepaid.)

☐ Check or Money Order Enclosed

☐ Visa ☐ Mastercard ☐ American Express

Account No. ____________________ Exp. Date __________

Signature ____________________

Shipping & Handling:

Amount of Order	United States	Canada & Mexico	All Others
Up to but not more than:	Add:	Add:	Add:
$20.00	$5.00*	$8.00	50% of amount
30.00	5.50	8.00	40% of amount
40.00	6.00	8.00	
50.00	6.50	14.00	
60.00	7.00	14.00	
80.00	8.00	14.00	
100.00	10.00	21.00	
150.00	12.50	28.00	
200.00	15.00	35.00	
More than $200.00	Add 20% of order	Add 20% of order	

*minimum amount for all orders

Note: Please make checks payable to the Water Environment Research Foundation.

To Order (Subscribers Only):

Log on to www.werf.org and click on the "Product Catalog."

Phone: (703) 684-2470
Fax: (703) 299-0742.

WERF
Attn: Subscriber Services
635 Slaters Lane
Alexandria, VA 22314-1177

To Order (Non-Subscribers):

Non-subscribers may be able to order WERF publications either through WEF (www.wef.org) or IWAP (www.iwapublishing.com). Visit WERF's website at www.werf.org for details.

IWA Publishing's authorised EU representative for General Product Safety Regulations is Diane D'Arras, 15 rue Duret, 75116 Paris, France, e-mail: safety@iwap.co.uk.

Printed and bound by CPI Group (UK) Ltd, Croydon, CR0 4YY

12/05/2026

02108881-0002